CONTENTS

Self-Esteem/Pride.........................2

Compassion.................................7

Courtesy 14

Responsibility 25

Fairness.....................................32

Citizenship.................................42

Good Character Activities............51

Word Cards 65

Character Cards 68

Student Awards and Certificates ..74

Rubrics 78

Character Education Book List.....80

Encouraging Interest

Help students to develop an understanding and appreciation for character education themes through reading stories. A list of picture books is included in this teacher resource.

Blackline Masters and Graphic Organizers

Encourage students to use the blackline masters and graphic organizers to present information, reinforce important concepts, and to extend opportunities for learning. The graphic organizers will help students focus on important ideas and make direct comparisons.

Character Cards

Use the provided character cards as a springboard for discussions and role playing, or to sort according to different character traits shown. You may also wish to enlarge the cards while photocopying and use them as a base for a Good Character bulletin board display.

Role Playing

Role playing offers an excellent opportunity for students to become sensitive to how others feel in different situations and to develop empathy. Be sure to introduce role playing only after class members become familiar and comfortable with each other. In addition, set rules for role playing to prevent inappropriate behavior. In order for students to get the most out of role playing, include:

- an enactment of the scenario presented
- a discussion and analysis of the scenario presented
- further role playing of alternatives
- drawing conclusions regarding the scenario presented

Character Education Activities 2–3 (USA Edition)

SELF-ESTEEM / PRIDE

Self-Esteem: confidence in your own worth or abilities; self-respect
Pride: delight or satisfaction in your accomplishments, achievements, and status

Activity 1: Student of the Week

The Student of the Week activity is not only a great way to promote self-esteem and to instill pride, but will encourage students to learn more about their classmates and to create a community. At the beginning of the school year, have families choose the week that their child will be Student of the Week. In preparation for that week ask families to send in special photos such as baby photos to display, and a bag of items that the student would like to share. Dedicate a bulletin board display with the student's information, pictures, and school work. The teacher may also wish to include written notes from the other students that compliment or recognize the child chosen for Student of the Week.

Activity 2: Celebrating Students

Acknowledge and celebrate children's accomplishments and positive qualities on an ongoing basis using the various certificates provided in this teacher resource. Keep track of which certificates have been handed to whom in order to watch out for certain behaviors or accomplishments for certain students. Certificates can be given out in the moment, or you may wish to hold a regular class meeting to recognize students.

Activity 3: Perseverance

Have students set personal goals. Encourage students to persevere and to achieve their goals:

• Affirm to students your confidence that they can achieve their goals.
• Give students honest feedback on what they are doing well and what they need to work on.
• If a task seems overwhelming to a student, break it down into smaller, more manageable parts.
• Let students know that it is okay if something is not easy, and that they can work through obstacles.
• Stress the importance of "finishing what you have started."
• Talk about your own experiences.
• Celebrate accomplishments and have students express how they feel when they have achieved a goal.

Activity 4: Developing Good Work Habits

Help students take responsibility for their learning. Encourage students to self-assess their daily work habits using kid-friendly criteria that are easy to understand. The How Am I Doing? rubric provided in this teacher resource will clarify what makes a good piece of work exemplary and the qualities of an excellent student.

_____ 'S GOAL

To achieve this goal, I need to

_____

_____

I want to achieve this goal because

_____

_____

My goal is to

_____

_____

Complete the sentence inside the balloon.

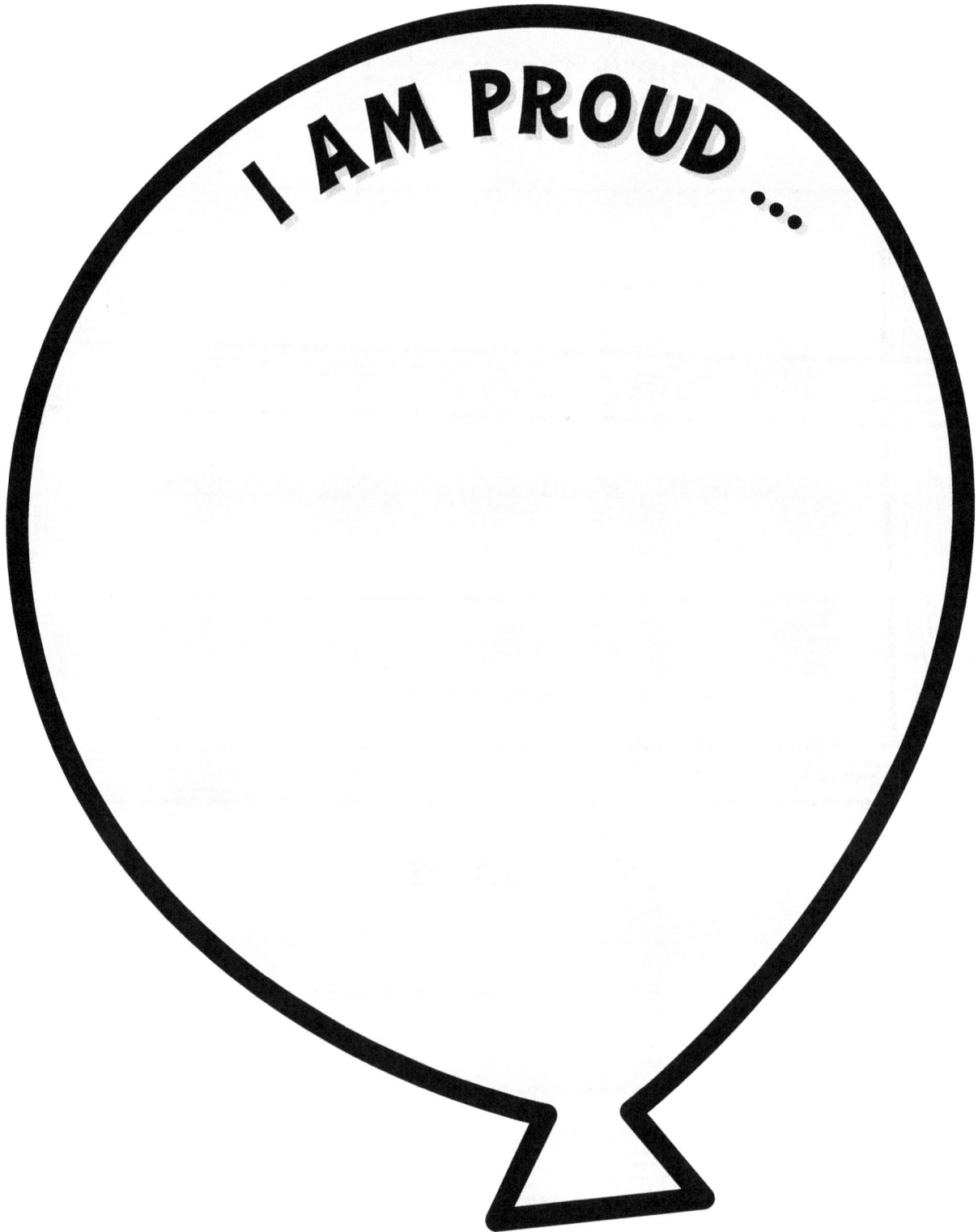

I AM PROUD ...

Student Of The Week: _____

Dear Parent/Guardian,

Your child has been chosen to be *Student of the Week* for the week of _____.
Please place in the paper bag special items from home that your child would like to bring back to share with our class. Be sure to include photos of your child's choice to display on our *Student of the Week* bulletin board display. In addition, please complete the information below to add to the display.

Your family's participation and support are greatly appreciated!

My favorite book:
_____

My favorite food:
_____

The best thing about school is
_____

WORK HABITS SURVEY

Good work habits help people get their work done. Here are some examples of good work habits. Take this survey and think about your work habits.

	Always	Sometimes	Never
I complete my work on time and with care.			
I use my time wisely.			
I follow directions.			
I keep my materials organized.			

Do you think you have good work habits? Explain your thinking.

_____

_____

_____

_____

_____

COMPASSION

Compassion: Sympathy and understanding toward the needs and feelings of others

Activity 1: People Have Feelings

Generate a class list of different kinds of feelings. Discuss situations that might occur around each feeling. Have students complete the My Feelings worksheets and discuss.

Activity 2: Caring People

Ask students, "What does it mean to be a caring person?" As a whole group brainstorm a list of do's and don'ts for being caring. Ask for specific examples of each behavior they identify.

Discussion Starters:
1. How do you think a new student feels when coming to a new class?
   What could you do?
2. What could you do to make a sad person happy?

Activity 3: Acts Of Kindness

Brainstorm with students what it means to be kind. Record their responses on chart paper. Next go through the student generated list and have students associate the kind of feelings they have around each act of kindness. Encourage students to understand that they have the ability to make someone happy, perhaps by complimenting them or doing something kind. Have students make compliment or appreciation cards for students in the class or create coupons to give out to people as an act of kindness.

Discussion Starters:
1. How does it feel to be kind? How does it feel to be mean?

Activity 4: When People Feel Angry...

Ask students to remember a time when they felt angry. Have students explain what happened and how they handled the situation. Some situations might include:

- something was unfair
- someone was mean or teased us
- something was broken
- someone was in our space
- someone was not sharing
- something was taken away from us.

Activity 5: Bullying

Help students gain a clear understanding of bullying. Bullying can be described as the act of hurting someone physically or psychologically. Students should also be made aware that bullies come in all shapes and sizes. Usually someone is bullied repeatedly. Some forms of bullying include:

Physical: hitting, punching, tripping, shoving, stealing belongings, locking someone in or out, etc.
Verbal: teasing, putdowns, taunting, making embarrassing remarks, etc.
Relational: excluding someone from a group, spreading rumors, ignoring someone, etc.

It is the hope that if students can understand what a person feels like when bullied, students will develop empathy and help stop bullying.

HOW WOULD YOU FEEL?

Describe how you would feel in each situation.

happy	sad	angry	worried	scared

Situation	I would feel ...
I am going to visit my favorite cousins.	
My pet fish died.	
I am moving to another school.	
It's my birthday.	
I had a test.	
Someone bullied me at school.	
My friend told me a funny joke.	
I had to try something for the first time.	
A special event got canceled.	

ACTS OF KINDNESS

Acts of kindness let people know that you care about them.
Color the shapes that are examples of acts of kindness.

listening

sharing your snack

being bossy

using manners

cooperating with others

including someone in a group

being helpful

being rude

teasing someone

I APPRECIATE YOU!

Thank you for...

_____

_____

_____

_____

_____

_____

_____

WHAT IS BULLYING?

Bullying is when someone mistreats someone on purpose, for example:

- name-calling or putdowns
- pushing or hitting
- ignoring or excluding

What are 3 things a person who is being bullied might do?

1. _____

2. _____

3. _____

What are 2 things you can do if you see someone being bullied?

1. _____

2. _____

PROBLEM REPORT

What happened?

_____

_____

_____

Has this ever happened to you before?   ☐ Yes   ☐ No

What did you do?

_____

_____

_____

If this happens, I should:

- Tell the person to stop
- Go to a safe place
- Tell a teacher

Other

_____

_____

A LETTER OF ADVICE

Choose:
- Write a letter of advice to someone who is being bullied.
- Write a letter of advice to someone who is being a bully.

Dear _____

_____

_____

_____

_____

_____

_____

_____

_____

_____

Your Friend,

_____

COURTESY

Courtesy: polite, considerate behavior toward others

Activity 1: Courtesy

Ask students if they know what the word *courtesy* or *polite* means. As a class, brainstorm a list of courtesy do's and don'ts. Create a class big book based on the list generated by the students.

Discussion Starters:

1. Why is it important to be polite to other people?
2. How do you feel when someone is polite to you?
3. How do you feel when you are polite?
4. How do you think others feel when you are polite to them?
5. Can you think of examples of how you can be polite to others today?

Activity 2: Encouraging Respect

Ask students what it means to treat other people with respect. Generate a class list of do's and don'ts for treating people with respect in different situations, such as when there is a class visitor. Post the list up on a wall as a reminder for students. Some of the do's and don'ts may include: DO be courteous and polite, DO listen to others without interrupting, DO treat other people the way you want to be treated, DON'T give people putdowns or treat them badly, and DON'T judge people before you get to know them.

Activity 3: People Are Alike

Encourage students to think about how people can be alike but still unique. Survey students on a variety of topics and create whole class graphs to demonstrate how people can be similar and/or different. Some survey ideas include: birthday months, favorite colors, number of people in their household, and favorite foods. In addition, celebrate the differences that students have.

Discussion Starters:

1. What do you notice?
2. What surprised you?

Activity 4: Friendship

Ask students to define friendship and if they think they have to be a good friend to have a good friend. Create a class "recipe" of behaviors for being a good friend. Discuss each one and have students name classmates who demonstrate each behavior. Some behaviors might be someone who: shares, is helpful, is kind, is fair, is fun, or is a good sport.

Discussion Starters:

1. I think the friendship behavior I am best at is …
2. I think the friendship behavior I need to work on is …

_____'s

BOOK ABOUT COURTESY

Courtesy is when you wait for your turn to speak, because ...

_____

_____

_____

Courtesy is when you use a tissue, because ...

_____

_____

_____

Courtesy is when you don't interrupt when someone is busy, because ...

_____

_____

_____

Courtesy is when you say "Thank You", because ...

_____

_____

_____

Courtesy is when you use good table manners, because ...

_____

_____

_____

COURTESY SURVEY

People get along better when they are courteous with each other. Here are some ways that people can be courteous with each other. Take the survey and think about how courteous you are with others.

	Always	Sometimes	Never
I wait for my turn to speak.			
I use a tissue when I sneeze.			
I wait my turn to speak, without interrupting.			
I use polite words.			
I use good table manners.			

Do you think you are a courteous person? Explain your thinking.

_____

_____

_____

_____

_____'s

BOOK ABOUT FRIENDS

What is a friend?

_____

_____

_____

19

Who is your best friend?

_____

Why is this person your best friend?

_____

_____

_____

What qualities do you have that make you a good friend?

_____

_____

_____

List 3 things you like to do with your friend.

_____

_____

_____

How are you and your friend alike?

_____

How are you and your friend different?

_____

FRIENDSHIP POEM

Write an acrostic poem about friends.

F ____________________

R ____________________

I ____________________

E ____________________

N ____________________

D ____________________

S ____________________

Design a t-shirt that gives a tip on how to be a good friend.

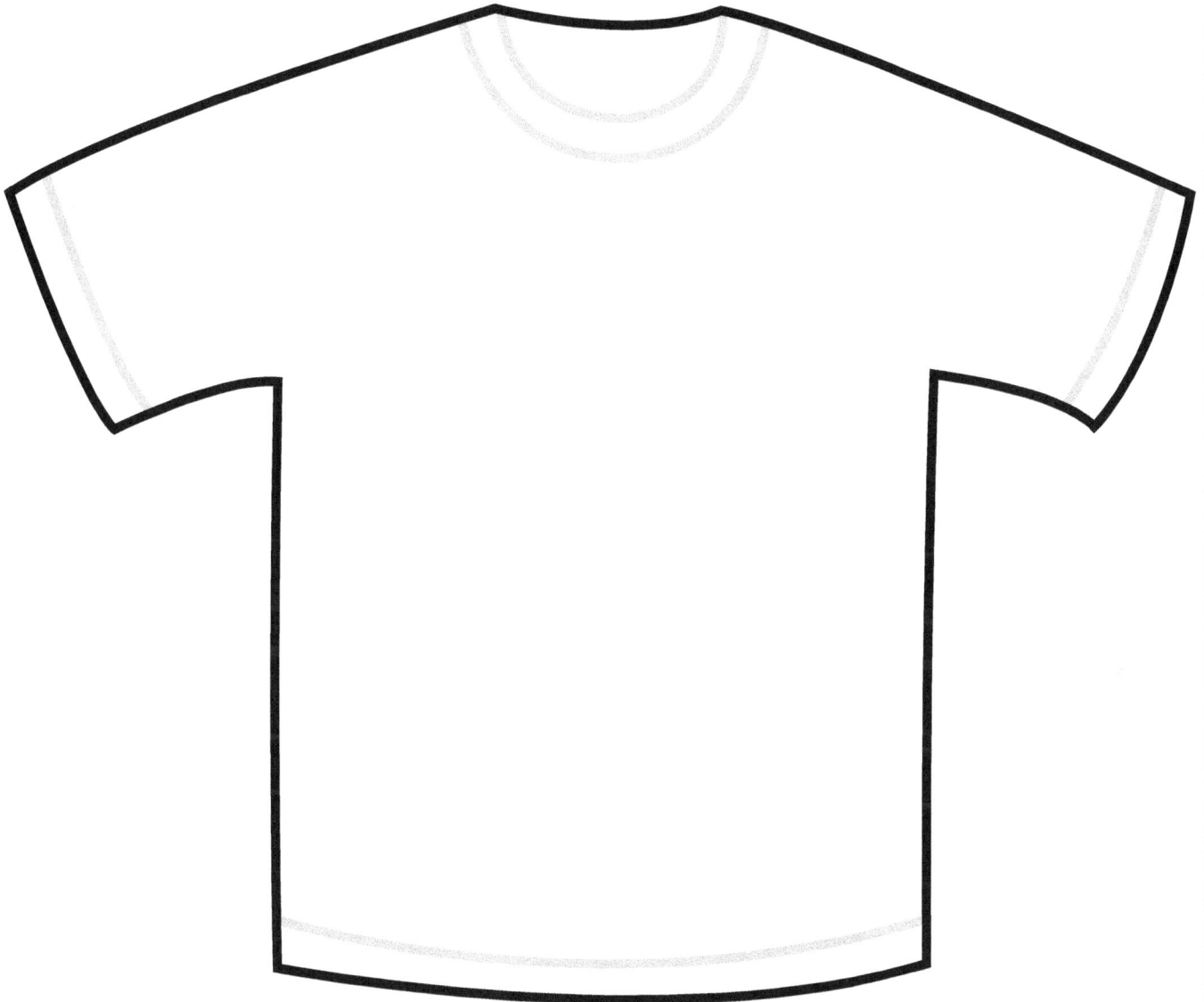

FRIENDSHIP SURVEY

Friendship skills are very important. Here are some ways you can show you're a good friend. Take this survey and think about your friendship skills.

	Always	Sometimes	Never
I share with my friends.			
I take turns.			
I help my friends.			
I have fun with my friends.			
I am a good listener.			

Do you think you are a good friend? Explain your thinking.

_____

_____

_____

_____

RESPONSIBILITY

A *responsibility* is a duty or task to be carried out carefully and thoroughly.
A *responsible* person is one that others can depend on and trust.

Activity 1: Families Work Together

As a whole group, ask students if they think families are important and to explain their thinking with examples. Then ask students to think about special contributions each family member makes to the family. Record the student responses on a chart. Put check marks or tally marks to show repeat answers. Encourage students to think about their own role in their family. What do they contribute? How is it helpful to the family?
Give students a family chart to take home and record how their family works together.

Activity 2: Rules and Responsibilities at Home

As a whole group, brainstorm different rules that students have at home. List the different rules on a chart and poll the students if they have each rule at their home. Some of the rules that might be stated are: I have a bedtime, I am not allowed to go near the stove without an adult, I need to be polite, I need to tidy up my toys, etc.

Discussion Starters:

1. Which rules keep you safe at home?
2. Which rules help keep you healthy?
3. Which rules help the members of your family to get along?
4. What do you think would happen if you didn't have any rules at home?
5. What rules would you change? Why?

Activity 3: Rules and Responsibilities at School

Using paper strips, have the students brainstorm classroom or school rules together. Some rules might include: walk in the hallways, keep your hands to yourself, be polite, ask permission to go to the bathroom, etc.

Discussion Starters:

1. Which rule do you think is the most important?
2. Who do you think should make up the rules in the classroom or at school? Explain your thinking.
3. Which rules help keep you safe?
4. Which rules help you learn?
5. What are your responsibilities at school?
6. What are the responsibilities of the people who work at your school?
7. Are there different rules for inside the school and outside the school?

Activity 4: Rules in Public Places

Do the same as above, but while talking about public places.

RULES ARE IMPORTANT

A rule at _____

_____

This rule is important because

_____

_____

_____ 's

Book About Responsibility

What does responsibility mean?

_____

_____

_____

In what ways do you show you're responsible at home?

_____

_____

_____

In what ways do you show you're responsible at school?

_____

_____

_____

In what ways do you show you're responsible when you're out in the community?

_____

_____

Why is it important to behave responsibly?

_____

_____

_____

Do you think it is important to take responsibility for your own actions? Tell why.

_____

_____

_____

What are the benefits of being a responsible person?

_____

_____

_____

RESPONSIBILITY SURVEY

Responsibility shows that you are dependable. Here are some ways you can show responsibility. Take this survey and think about how responsible you are.

	Always	Sometimes	Never
I take responsibility for my own actions.			
When I agree to do something, I do it.			
I show I am responsible by completing my chores.			
I show I am responsible by following the rules.			
I show I am responsible by getting my school work done.			

Do you think you are a responsible person? Explain your thinking.

_____

_____

_____

_____

FAIRNESS

Fair: treating everyone, including yourself, with the same strictness or kindness; also, following the rules

Activity 1: What is Fairness?

Discussion Starters:

1. What does treating people fairly mean?
2. Have you ever said, "That's unfair"?
   How do you know when something is unfair?
3. Have you ever played a game when someone cheated? How did you feel about it?
4. Does fairness mean enforcing the same rules for everyone, even if it means losing a game?

Activity 2: Honesty is the Best Policy

As a whole group ask students what they think the expression, "honesty is the best policy" means. Do they agree with the expression? Have students explain their thinking.

Discussion Starters:

1. Would you trust somebody who lies? Who cheats? Who steals? Why, or why not?
2. Have you ever told the truth when it was a difficult thing to do? Explain.

Activity 3: Making Good Decisions

Encourage children to learn to think about whether something is right or wrong before making a choice of how they will proceed in different situations. Role play the different scenarios found on the Good Character cards in this book. Compare and discuss what happens in each scenario with children who choose to do the "right thing" and with children who choose to do the "wrong thing." How would children feel after each decision? What are the consequences?

Discussion Starters:

1. What can you think about before deciding if doing something is right or wrong?
2. What do you think would happen if nobody cared about doing the "right thing"?
3. Do you agree with "finders keepers, losers weepers"? Explain your thinking.

Activity 4: What is Conflict Resolution?

Introduce the idea of conflict resolution to students. Conflict resolution is a process to help solve problems in a positive way. Each person involved is encouraged to take responsibility for their actions. For younger children you may wish to refer to it as "working it out."
Clear steps for conflict resolution might include:

- Identify the problem.
- Listen without interrupting.
- Talk it out.
- Come up with different solutions.

Discuss and review the above process with students. Role play different situations so students can practise walking through the process. Students should be encouraged to try to understand the other person's perspective of a conflict. You may wish to use situations that are present in your class. Encourage students to think of different solutions so they have options if one solution does not work. In addition, post the steps for conflict resolution on the board for easy student reference.

DISCUSSION STARTER CARDS

GOOD CHARACTER DISCUSSION STARTER

One of your classmates needs a pencil, but doesn't have one. You have an extra pencil.

What could you do?

GOOD CHARACTER DISCUSSION STARTER

You see someone on the schoolyard who fell and is crying.

What would you do?

GOOD CHARACTER DISCUSSION STARTER

Name a fair way to decide who goes first in a game.

Explain your reasons.

GOOD CHARACTER DISCUSSION STARTER

You are lining up for lunch and see someone's money fall out of their pocket.

What is the right thing to do?

GOOD CHARACTER DISCUSSION STARTER

You are really upset that someone has taken your favorite pencil off your desk.

What would you do?

GOOD CHARACTER DISCUSSION STARTER

You see someone being bullied in the schoolyard.

What is the right thing to do?

DISCUSSION STARTER CARDS

GOOD CHARACTER DISCUSSION STARTER

There is a new student in your class. What would you do to make them feel welcome?

GOOD CHARACTER DISCUSSION STARTER

There is a special toy you would like to buy but you don't have enough money right now.

What can you do?

GOOD CHARACTER DISCUSSION STARTER

You want to get your caregiver's attention, but they are on the phone.

What should you do?

GOOD CHARACTER DISCUSSION STARTER

Your mother wants you to get ready for bedtime, but you don't want to.

What should you do?

GOOD CHARACTER DISCUSSION STARTER

You have just spilled something at home and have made a big mess.

What should you do?

GOOD CHARACTER DISCUSSION STARTER

How can you show that you use your time wisely at school?

Explain your answers.

_____'S

BOOK ABOUT FAIRNESS

What is fairness?

_____

_____

_____

Why do you think it's important to treat people the way you would want to be treated?

_____

_____

_____

Why do you think it's important to play by the rules?

_____

_____

_____

How do you know when something is unfair?

_____

_____

_____

What could you do when you are treated unfairly?

_____

_____

_____

WORK IT OUT!

What is the problem?

Listen without interrupting.

Talk it out.

Come up with different solutions.

FAIRNESS SURVEY

People get along better when they are fair with each other. Here are some ways that people can be fair with each other. Take the survey and think about how fair you are with others.

	Always	Sometimes	Never
I show I am fair by taking turns.			
I treat people the way I want to be treated.			
I take responsibility for my share of group work.			
I tell people when they are doing something well.			
I talk about disagreements and look for a solution.			

Do you think you are a fair person? Explain your thinking.

_____

_____

_____

_____

CITIZENSHIP

Citizenship: A person who is a good citizen is law abiding and involved in service to school community and country.

Activity 1: What is Citizenship?

Introduce the idea of citizenship to students in a whole group setting. Reinforce with the children that they all have something to contribute to the class, school, and community.

Brainstorm a list of useful things that children could do to help out in class, at home, in the school, and within the community.

Discussion Starters:

1. How do you feel after you have helped someone?
2. How do you feel after you have been helped?
3. Who are some people you know that volunteer at school or in the community? Why do you think they do it?
4. How does obeying the rules show that you are a good citizen?
5. How do community workers make the community a better place?

Activity 2: Making a Difference

As a whole group brainstorm a list of people who may be in need. Encourage students to think about people they know, or situations such as children in impoverished countries. In addition, discuss charities that children might be familiar with and any fundraising activities they have participated in at school or with their families.

Give the students in your class the opportunity to practise citizenship by taking part in a school or community project. Some ideas include:

- Planting a school garden
- Collecting toys for needy children
- Collecting clothes for a shelter
- Participating in a book drive for the school or shelter
- Collecting school supplies for children in impoverished countries
- Collecting change for a charity
- Participating in a trash pick-up day
- Going on a class field trip to a senior citizens' home to perform songs or to sit and play games with the residents

Activity 3: Citizenship Class Collage

Create a class citizenship collage using words, artwork, and pictures cut out from magazines or newspapers.

_____'s

BOOK ABOUT CITIZENSHIP

What does citizenship mean?

_____

_____

_____

_____

A good citizen is someone who takes responsibility for themselves.

List 2 examples.

_____

_____

A good citizen is someone who tries to make the world a better place.

List 2 examples.

_____

_____

A good citizen is someone who cares about the environment. What are some ways you can show you care about the environment?

_____

_____

_____

A good citizen is someone who follows the rules. What rules do you follow to show you're a good citizen?

_____

_____

_____

A good citizen is someone who takes part in the community.

List 2 examples.

_____

_____

A good citizen is someone who treats people with respect. How can you show others respect?

_____

_____

_____

CITIZENSHIP COLLAGE

Cut and paste pictures of ways you can show good citizenship.

Write about your collage.

_____

_____

_____

Design a citizenship crest to let people know how to be a good citizen.

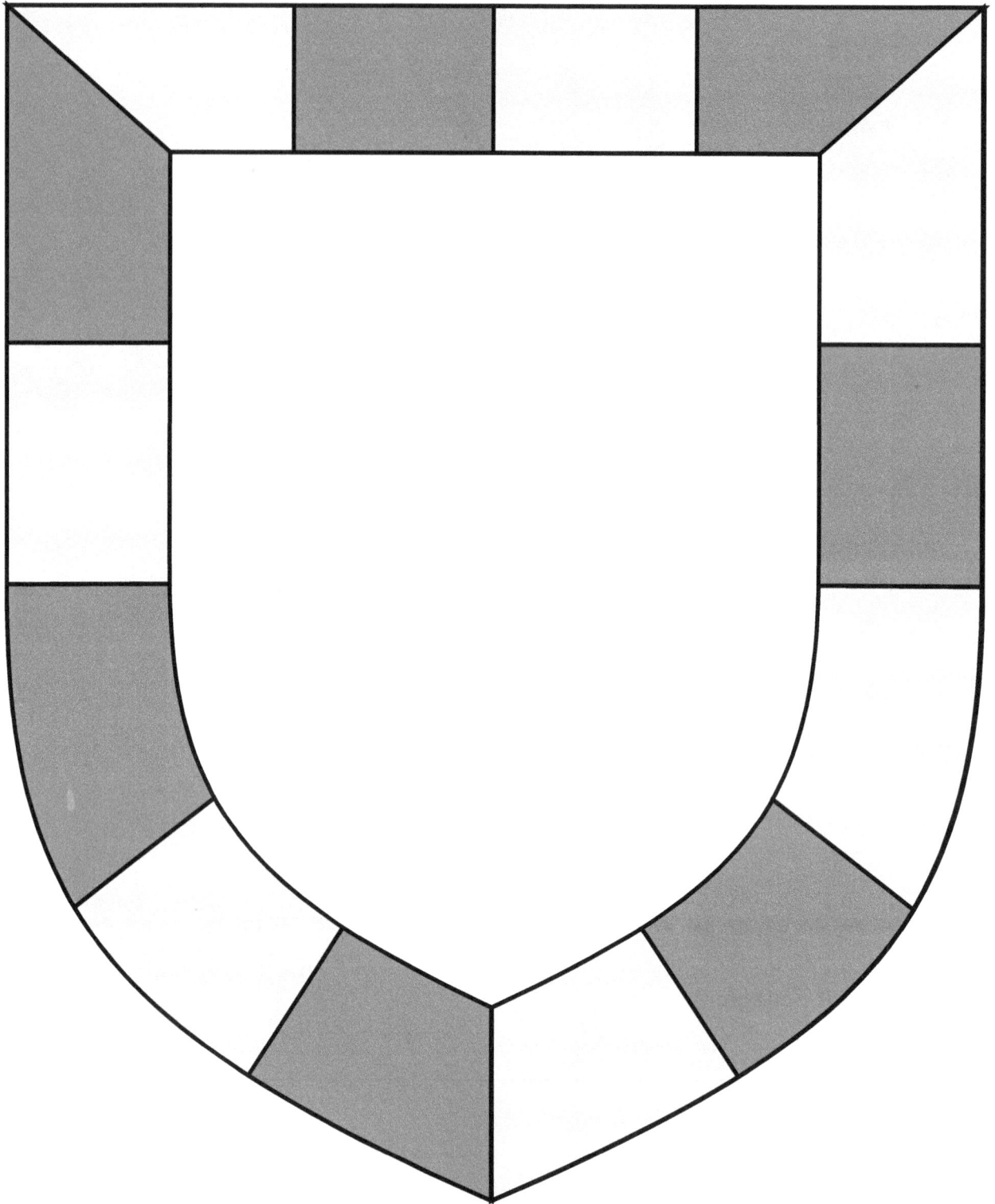

COOPERATION SURVEY

People get along better when they cooperate with each other. Here are some ways that people can cooperate with each other. Take the survey and think about how well you cooperate with others.

	Always	Sometimes	Never
I share things with others.			
I wait for my turn.			
I take responsibility for my share of group work.			
I tell people when they are doing something well.			
I talk about disagreements and look for a solution.			

Do you think you are a cooperative person? Explain your thinking.

_____

_____

_____

_____

CITIZENSHIP SURVEY

People get along better when everyone is a good citizen. Here are some ways that show how to be a good citizen. Take the survey and think about how you can be a good citizen.

	Always	Sometimes	Never
I try to make the world a better place.			
I care about the environment.			
I follow the rules.			
I take part in the community.			
I treat people with respect.			

Do you think you are a good citizen? Explain your thinking.

_____

_____

_____

_____

SOMEONE I ADMIRE

Draw a picture of someone you admire.

Write about why you admire this person.

_____

_____

_____

_____

_____

_____

_____

THINKING ABOUT GOOD CHARACTER

My Journal about:

_____

_____

_____

_____

_____

_____

_____

_____'s

GOOD CHARACTER DIARY

Ideas for your diary:

- Write about the ways you have been a good citizen.
- Write about the ways you have shown courtesy.
- Write about what you have done to achieve a special goal.
- Write about the ways you have shown you are responsible.
- Write about the ways you have shown you are a good friend.

GOOD CHARACTER DIARY

Monday

GOOD CHARACTER DIARY

Tuesday

GOOD CHARACTER DIARY

Wednesday

GOOD CHARACTER DIARY
Thursday

GOOD CHARACTER DIARY
Friday

GOOD CHARACTER DIARY

Saturday

GOOD CHARACTER DIARY

Sunday

Choose a character trait and create a poster about it.

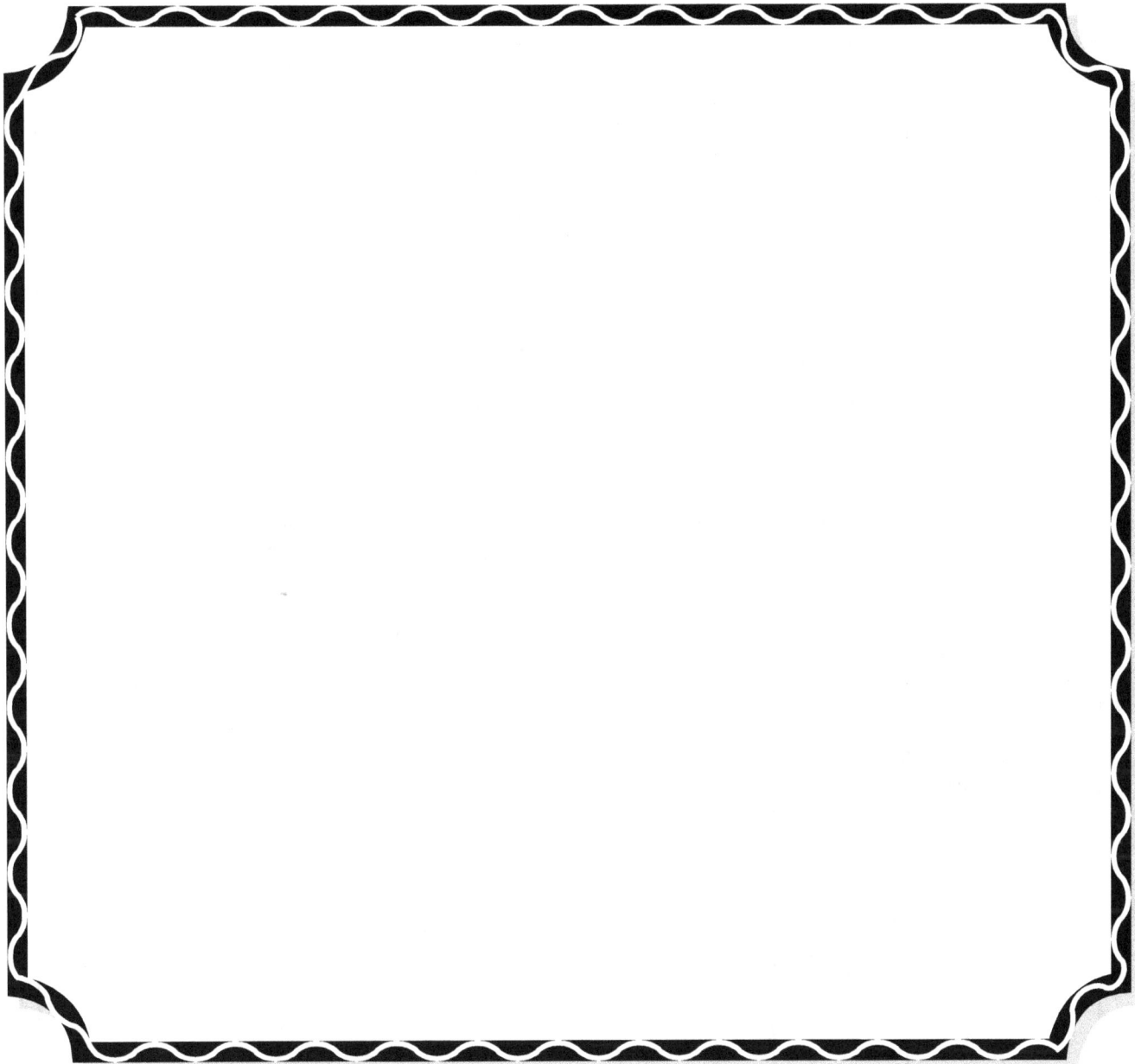

Write about your poster.

_____

_____

_____

GOOD CHARACTER WORD SEARCH

R	P	R	I	D	E	R	O	U	S	T
E	E	H	K	I	N	D	N	E	S	S
S	Z	O	P	T	G	A	I	C	P	C
P	Z	N	F	C	I	D	C	O	O	I
O	C	E	A	H	C	I	K	O	R	T
N	O	S	I	A	O	L	T	P	T	I
S	M	T	R	R	U	I	O	E	S	Z
I	P	Y	N	A	R	G	L	R	M	E
B	A	T	E	C	T	E	A	A	A	N
I	S	A	S	T	E	N	P	T	N	S
L	S	I	S	E	S	C	E	I	S	H
I	I	L	E	R	Y	E	B	O	H	I
T	O	B	O	I	G	R	A	N	I	P
Y	N	R	E	S	P	E	C	T	P	Y

character	courtesy	kindness
citizenship	diligence	pride
compassion	fairness	respect
cooperation	honesty	responsibility

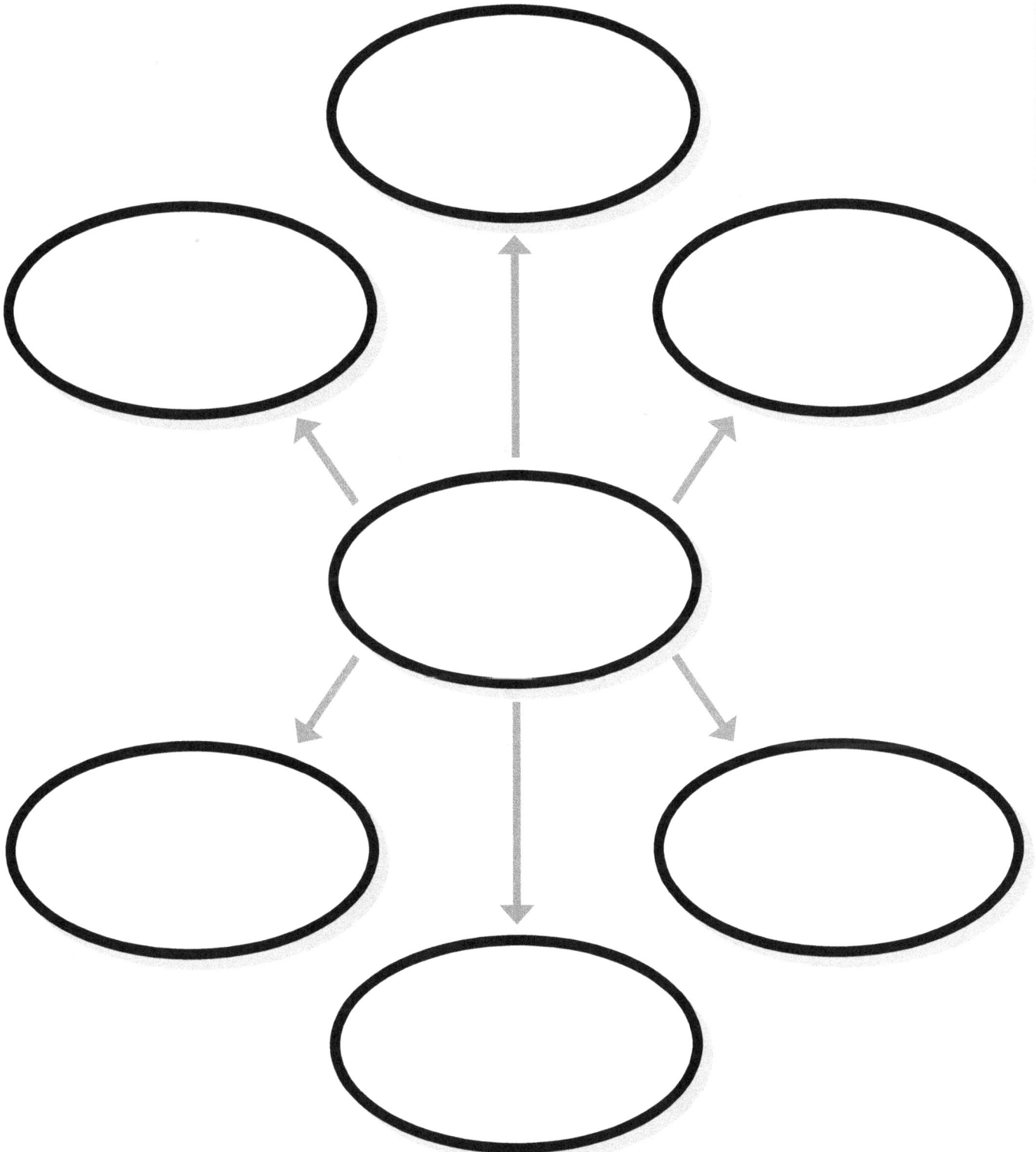

A T-CHART ABOUT...

_____

Write about your stamp.

_____

_____

Write a character trait in the center of the flower. In the petals, give examples of the character trait.

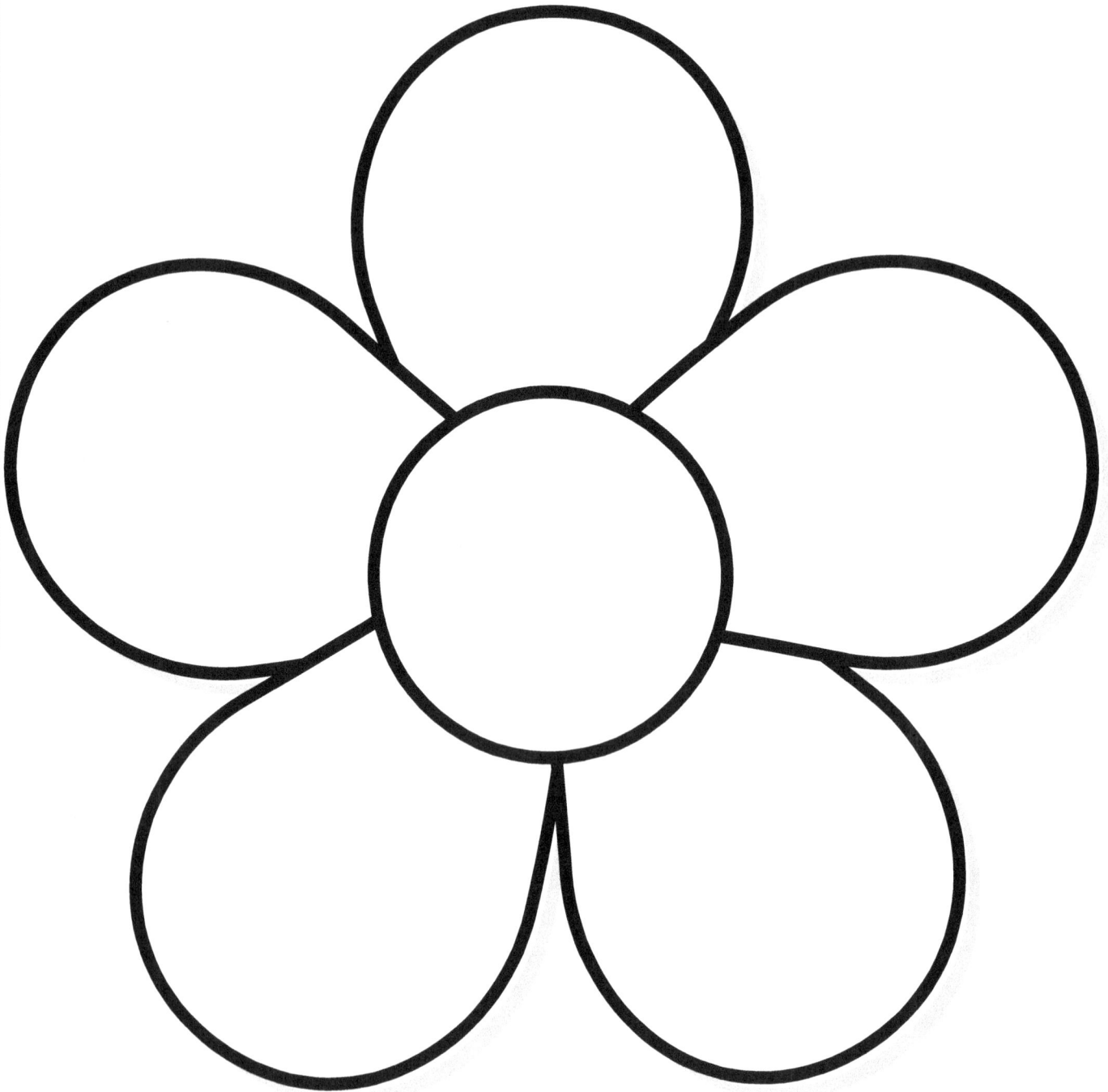

CLASS GOOD CHARACTER QUILT

1. In the quilt shape, draw a self-portrait and finish the sentence starter.
2. Carefully punch holes through the border around the inner square.
3. Using yarn, lace together the paper squares to form a class good character quilt.

I show good character by

_____

_____

Use this cube to display information about a character trait.

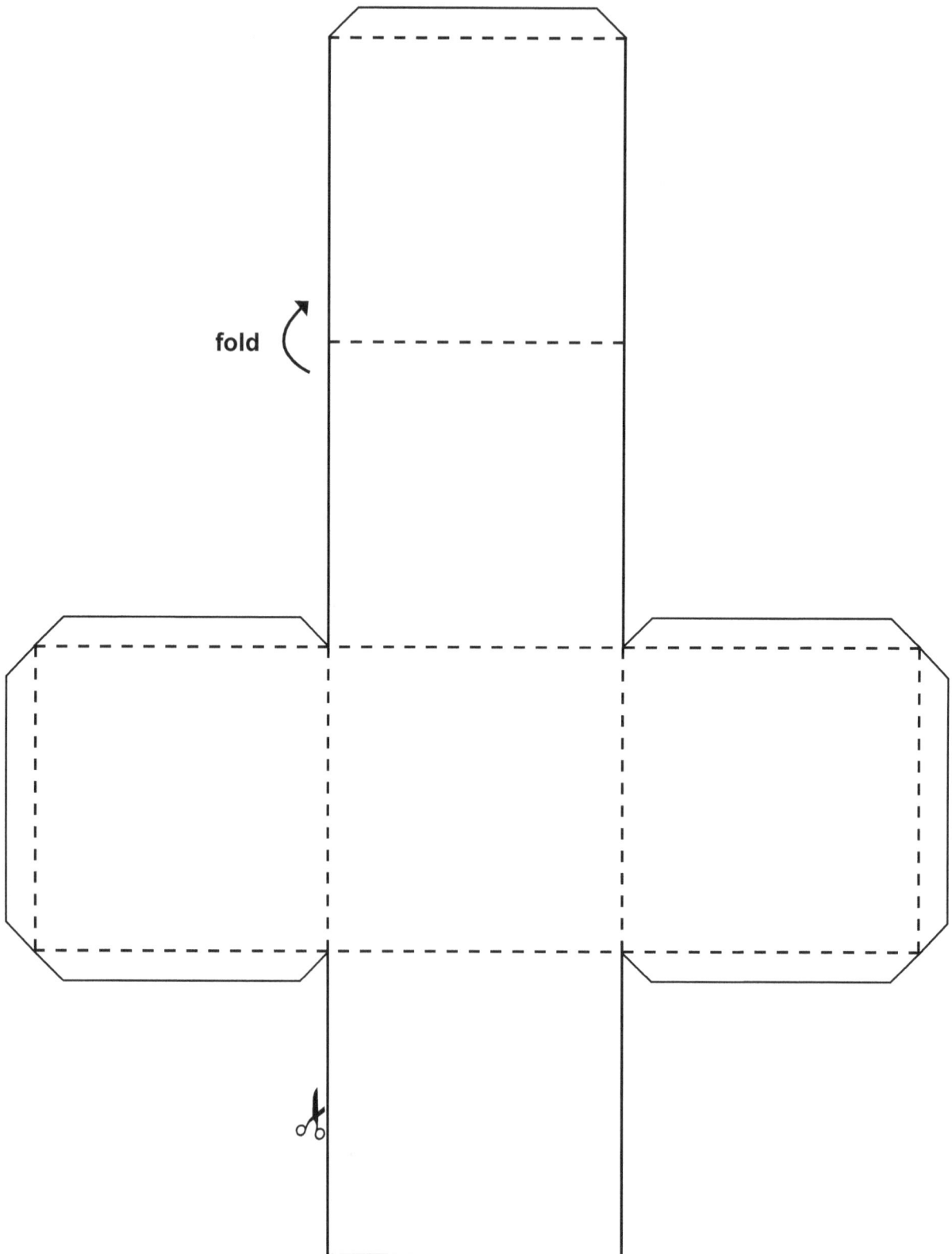

fold

responsibility

courtesy

65

compassion

fairness

citizenship

friendship

GOOD CHARACTER CARDS

May I have a treat?

GOOD CHARACTER CARDS

GOOD CHARACTER CARDS

GOOD CHARACTER CARDS

GOOD CHARACTER CARDS

GOOD CHARACTER CARDS

RESPONSIBILITY AWARD

Keep up the effort!

GREAT FRIEND AWARD

This award is for

_____

COURTESY AWARD

KEEP UP THE EFFORT!

TOP MANNERS!

THIS AWARD IS FOR

_____

GREAT JOB!

Keep up the effort!

SUPER
SPORTSMANSHIP!

This award is for

_____

CITIZENSHIP AWARD

KEEP UP THE EFFORT!

Community Helper AWARD

This award is for

_____

STUDENT PARTICIPATION RUBRIC

LEVEL	STUDENT PARTICIPATION DESCRIPTOR
Level 4	Student consistently contributes to class discussions and activities by offering ideas and asking questions.
Level 3	Student usually contributes to class discussions and activities by offering ideas and asking questions.
Level 2	Student sometimes contributes to class discussions and activities by offering ideas and asking questions.
Level 1	Student rarely contributes to class discussions and activities by offering ideas and asking questions.

UNDERSTANDING OF CONCEPTS RUBRIC

LEVEL	UNDERSTANDING OF CONCEPTS DESCRIPTOR
Level 4	Student shows a thorough understanding of all or almost all concepts and consistently gives appropriate and complete explanations independently. No teacher support is needed.
Level 3	Student shows a good understanding of most concepts and usually gives complete or nearly complete explanations. Infrequent teacher support is needed.
Level 2	Student shows a satisfactory understanding of most concepts and sometimes gives appropriate, but incomplete explanations. Teacher support is sometimes needed.
Level 1	Student shows little of understanding of concepts and rarely gives complete explanations. Intensive teacher support is needed.

COMMUNICATION OF CONCEPTS RUBRIC

LEVEL	COMMUNICATION OF CONCEPTS DESCRIPTOR
Level 4	Student consistently communicates with clarity and precision in written and oral work. Student consistently uses appropriate terminology and vocabulary.
Level 3	Student usually communicates with clarity and precision in written and oral work. Student usually uses appropriate terminology and vocabulary.
Level 2	Student sometimes communicates with clarity and precision in written and oral work. Student sometimes uses appropriate terminology and vocabulary.
Level 1	Student rarely communicates with clarity and precision in written and oral work. Student rarely uses appropriate terminology and vocabulary.

HOW AM I DOING?

	COMPLETING MY WORK	USING MY TIME WISELY	FOLLOWING DIRECTIONS	KEEPING ORGANIZED
FULL SPEED AHEAD!	• My work is consistently complete and done with care. • I add extra details to my work.	• I consistently get my work done on time.	• I consistently follow directions.	• My materials are consistently neatly organized. • I am consistently prepared and ready to learn.
KEEP GOING!	• My work is complete and done with care. • I check my work.	• I usually get my work done on time.	• I usually follow directions without reminders.	• I usually can find my materials. • I am usually prepared and ready to learn.
SLOW DOWN!	• My work is complete. • I need to check my work.	• I sometimes get my work done on time.	• I sometimes need reminders to follow directions.	• I sometimes need time to find my materials. • I am sometimes prepared and ready to learn.
STOP!	• My work is not complete. • I need to check my work.	• I rarely get my work done on time.	• I need reminders to follow directions.	• I need to organize my materials. • I am rarely repared and ready to learn.

CHARACTER EDUCATION PICTURE BOOK LIST

ANTI-BULLYING

Oliver Button is a Sissy, written by Tomie dePaola
Benny Gets a Bully Ache, written by Jane Bomberge
Big Bad Bruce, written by Bill Peet
Martha Walks the Dog, written by Susan Meddaugh

COMPASSION

Berenstain Bears and the In-Crowd, written by Jan & Stan Berenstain
Frog and Toad Are Friends, written by Arnold Lobel
A Chair For My Mother, written by Vera B. Williams

FRIENDSHIP

Alexander and the Windup Mouse, written by Leo Lionni
Best Friends, written by Steven Kellogg

HONESTY / FAIRNESS

It's Not Fair!, written by Charlotte Zolotow
A Little Princess, written by Frances Hodgson Burnett
Alexander And The Terrible, Horrible, No Good, Very Bad Day, written by Judith Viorst

MANNERS

Manners, written by Aliki
What Do You Do Dear?, written by Seysle Joslin

PERSEVERANCE

A Weed Is A Flower, written by Aliki
Green Eggs and Ham, written by Dr. Seuss
Mike Mulligan and His Steam Shovel, written by Virginia Lee Burton
The Very Busy Spider, written by Eric Carle
The Little Engine That Could, written by Watty Piper

PRIDE

The Summer of the Swans, written by Betsy Byars
Captain Tom Cat, written by Bill Martin Jr.
The Trumpet of the Swan, written by E. B. White

RESPECT

Goldilocks and the Three Bears, written by various authors
Grandfather Counts, written by Andrea Cheng

RESPONSIBILITY

Berlioz the Bear, written by Jan Brett
Now One Foot, Now the Other, written by Tomie dePaola
Horton Hatches the Egg, written by Dr. Seuss
Strega Nona: An Old Tale, written by Tomie dePaola

www.ingramcontent.com/pod-product-compliance
Lightning Source LLC
Chambersburg PA
CBHW081205270326
41930CB00014B/3315